Mathematical Foundation of the Quantum Theory of Gravity

Beyond Einstein, Volume 3

Balungi Francis

Published by Bill Stone Services, 2018.

Also by Balungi Francis

Beyond Einstein
Quantum Gravity in a Nutshell1
Solutions to the Unsolved Physics Problems
Mathematical Foundation of the Quantum Theory of Gravity
A New Approach to Quantum Gravity
Balungi's Approach to Quantum Gravity
QG: The strange theory of Space,Time and Matter
The Holy Grail of Modern Physics
Fifty Formulas that Changed the World
Quantum Gravity in a Nutshell1 Second Edition
What is Real?:Space Time Singularities or Quantum Black Holes?Dark Matter
or Planck Mass Particles? General Relativity or Quantum Gravity? Volume or
Area Entropy Law?
The Holy Grail of Modern Physics
Brief Solutions to the Big Problems in Physics, Astrophysics and Cosmology

Brief Solutions to the Big Problems
Brief Solutions to the Big Problems

Pursuing a Nobel Prize
Serious Scientific Answers to Millennium Physics Questions

Using Geographical Information Systems to Create an Agroclimatic Zone map for Soroti District

Think Physics
Proof of the Proton Radius
Emergence of Gravity
On the Deflection of Light in the Sun's Gravitational Field
Reinventing Gravity

Standalone
Using Gis to Create an Agroclimatic Zone map for Soroti Distric
Expecting
Quantum Gravity in a Nutshell 2
Balungi's Guide to a Healthy Pregnancy
Prove Physics
The Origin of Gravity and the Laws of Physics
Derivation of Newton's Law of Gravitation
When Gravity Breaks Down

Table of Contents

.... 1

Dedication 2

PREFACE.. 3

Introduction 4

CHAPTER1:The Study Of Gravitational And Electromagnetic Radiation Intensity On a Quantum Scale as a Basis For the Development of the Theory of Quantum Gravity.. 9

CHAPTER2:The Magnetic Fields at Which the Theories of Quantum Electrodynamics and Quantum Gravity Become Analogous (Field limit) 14

CHAPTER3:Derivation of the temperature and entropy of black holes... 16

CHAPTER4:The Art of Reductionism.. 25

CHAPTER5:On the Bohr's model of the Hydrogen atom from a Quantum Gravity perspective and the deduction of the earliest period of time 27

CHAPTER6:Deduction of the maximal magnetic field, radiation intensity, quantum hall effect and the laws of black hole mechanics from a proposed theory of quantum gravity 31

CHAPTER7: On The Complete Theory of Light 36

CHAPTER8:Construction of a Consistent Physical Theory of Nature 43

CHAPTER9:The Basic Wave Equation (Derivation of the Fermi's Energy).. 49

References 50

Mathematical Foundation of a Quantum Theory of Gravity

BALUNGI FRANCIS

Visionary School of Quantum Gravity
+256(0)777105605
Email: balungif@gmail.com
First Book Edition: 2016

Dedication

To my wife for his constant feedback throughout and many long hours of editing, and friends who offered their time and comments along the way

PREFACE

There is a need for a book on a Quantum Theory of Gravity that is not directed at specialists but, rather, sets out the concepts underlying this subject for a broader scientific audience and conveys joy in their beauty. Balungi has written with this goal in mind, and has succeeded admirably. This wonderful and exciting book is optimal for physics graduate students and researchers. The physical explanations are exceedingly well written and integrated with formulas. Quantum Gravity is the next big thing and this book will help the reader to understand and use the theory

Balungi Francis

Introduction

The development of a quantum theory of gravity began in 1899 with Max Planck's formulation of "Planck scales" of mass, time, and length. During this period the theories of quantum mechanics, quantum field theory and general relativity had not yet been developed, which means that Planck himself had no idea about what he had just developed, "Behind the Black board". Planck was not aware of quantum gravity and what it would mean for physicists. But he had just coined in formula one of the starting point for the holy grail of physics.

After P.Bridgman's disapproval of Planck's units in 1922, Albert Einstein having published the General Relativity theory, a few months after it's publication he noted that *"to the intra-atomic movement of electrons, atoms would have to radiate not only electromagnetic but also gravitational energy if only in tiny amounts. As this is hardly true in nature, it appears that quantum theory would have to modify not only Maxwellian electrodynamics, but also the new theory of gravitation"*. This showed Einstein's interest in the unification of Planck's quantum theory with his newly developed theory of Gravitation.

Then in 1933 came Bronstein's cGh-plan as we know it today. In his plan he argued a need for Quantum Gravity in this statement *"After the relativistic quantum theory is created, the task will be to develop the next part of our scheme that is, to unify quantum theory (h), special relativity (c) and the theory of gravitation (G) into a single theory".* Thus the theory of quantum gravity is expected to be able to provide a satisfactory description of the microstructure of space time at the so called Planck scales, at which all fundamental constants of the ingredient theories, c (speed of light), h (Planck constant) and G (Newton's constant), come together to form units of mass, length and time.

Therefore the need for the theory of quantum gravity is crucial in understanding nature, from the smallest to the biggest particle ever known in the universe, for example "we can describe the behavior of flowing water with the long- known classical theory of hydrodynamics, but if we advance to smaller and smaller scales and eventually come across individual atoms, it no longer applies. Then we need quantum physics just as a liquid consists of atoms" (Daniel Oriti), he imagines space to be made up of tiny cells or atoms of space, and a new theory of quantum gravity is required to describe them fully".

The aim of this book is to develop a theory capable of explaining the quantum behavior of the gravitational fields and thereafter explain the problems involving the combination of very high energy and very small dimensions of space, such as the behavior of Black holes and the study of the properties of the early universe.

One of the major problems in developing a quantum theory of gravity lies in calculating and finding an expression for the precise acceleration of particles on a quantum scale. The problem is important in calculating various limiting cases for both the theory of quantum gravity (at the Planck epoch) and the theory of quantum electrodynamics. As we shall see, all the calculations undertaken in the process lead us to one thing, a consistent quantum theory of gravity. I am not sure whether I have deduced in principle the quantum theory of gravity but I will leave it to the reader and the entire physics community to decide. To differ from Newton's laws of motion and Einstein's theory of general relativity, our acceleration will depend on the dimensionless coupling constant which determines the strength of the force in any given interaction as,

$$a_{acel} = \frac{c^4}{e}\left(\frac{4\pi\varepsilon_0\alpha}{G}\right)^{1/2}$$

Where, c is the constant speed of light, e is the charge on an electron, α is the dimensionless coupling constant, ε is the permittivity of free space and G is the universal gravitational constant.

Various examples have been given below in which the theories of quantum gravity and quantum electrodynamics will act as limiting cases,

(1)For example, where the quanta exchanged between two electrons is a photon in case of the electromagnetic force we have the electromagnetic coupling constant or the fine structure constant as, $\alpha = \frac{e^2}{4\pi\varepsilon_0\hbar c}$ which deduces the acceleration to, $a = \frac{c^{7/2}}{(\hbar G)^{1/2}}$. This is the allowed maximum acceleration for quantum gravitational effects at the Planck epoch. But for the case where the quanta exchanged between two electrons is a graviton for a gravitational force, we have the gravitational coupling constant as, $\alpha = \frac{Gm^2}{\hbar c}$ which gives the acceleration on a quantum electrodynamics scale as, $a = \frac{m}{e}\left(\frac{4\pi\varepsilon_0 c^7}{\hbar}\right)^{1/2}$.

(2)Then the minimal radius to which a gravitating body or an electron can collapse in a commoving frame can also be deduced as, If we equate Newton's law of universal gravitation to our newly developed force as, $\dfrac{Gm^2}{R^2} = \dfrac{mc^4}{e}\left(\dfrac{4\pi\varepsilon_0\alpha}{G}\right)^{1/2}$

We obtain the area as, $R^2 = \dfrac{me}{c^4}\left(\dfrac{G^3}{4\pi\varepsilon_0\alpha}\right)^{1/2}$ then for $\alpha = \dfrac{Gm^2}{\hbar c}$, as in the first example, we obtain the minimum radius for a charged particle for quantum gravitational effects as,

$$R_{min1} = (Ge)^{1/2}\left(\dfrac{\hbar}{4\pi\varepsilon_0 c^7}\right)^{1/4} = 4.717444838 \times 10^{-36} m,$$ or

$R_{min1} = 0.2923 l_p$, where l_p is the Planck length. Then the fine structure constant will be calculated as, $\alpha = \left(\dfrac{R_{min1}}{L_p}\right)^4 = \dfrac{1}{(3.42155)^4} = \dfrac{1}{137.054}$ Hence solving one of the unsolved problems in physics.

But for $\alpha = \dfrac{e^2}{4\pi\varepsilon_0\hbar c}$, we obtain the minimal radius due to torsion in the gravitational field as, $R_{min2} = (m)^{1/2}\left(\dfrac{\hbar G^3}{c^7}\right)^{1/4}$.

(3) On another note, we could derive the mass formula only if we equate the force $F = \dfrac{\hbar c^5}{G^2 m^2}$ to our force formula $F = \dfrac{mc^4}{e}\left(\dfrac{4\pi\varepsilon_0\alpha}{G}\right)^{1/2}$, then the mass expression is deduced as $m = \dfrac{\hbar^{1/3}e^{1/3}c^{1/3}}{(4\pi\varepsilon_0)^{1/6}G^{1/2}\alpha^{1/6}}$ This gives the Planck mass at $\alpha = \dfrac{e^2}{4\pi\varepsilon_0\hbar c}$, Also the mass that incorporates all the constants of nature when $\alpha = \dfrac{Gm^2}{\hbar c}$ is deduced as, $m = \left(\dfrac{(\hbar c)^3 e^2}{4\pi\varepsilon_0 G^4}\right)^{1/8} = 1.177535 \times 10^{-8} kg$. This could be the mass of the graviton.

(4)Then from Einstein's proposal for the radiation of the gravitational energy, we have an expression for energy as, $W = ma_{accel}R = \dfrac{mc^4}{e}\left(\dfrac{4\pi\varepsilon_0\alpha}{G}\right)^{1/2}R$, Where R is the radius of orbit of an electron around the nucleus of an atom, for $R \sim \dfrac{\hbar}{mc}$, and $\alpha = \dfrac{e^2}{4\pi\varepsilon_0\hbar c}$ we obtain the maximum energy as, $W = \left(\frac{\hbar c^5}{G}\right)^{1/2}$. This is the Planck energy at the Planck epoch.

But for $R \sim \frac{\hbar}{mc}$ and $\alpha = \frac{Gm^2}{\hbar c}$ we obtain the energy possessed by an electron

of mass m in the electromagnetic field as, $W = \frac{2m}{e}\left(\pi\varepsilon_0 c^5 \hbar\right)^{1/2}$, Then at the

Planck epoch when $m = \sqrt{\frac{\hbar c}{G}}$ (the Planck mass), the energy required to accelerate an electron in the gravitational field will be given by,

$W = \frac{2\hbar c^3}{e}\left(\frac{\pi\varepsilon_0}{G}\right)^{1/2}$ .In the case where the energy of the quantized states of the

hydrogen atom $\frac{me^4}{16\pi^2 n\hbar^2 \varepsilon_0^2}$, is equated to the energy of $\frac{2m}{e}\left(\pi\varepsilon_0 c^5 \hbar\right)^{1/2}$, we

obtain a crucial relationship between the fine structure constant and the

principal quantum number n, as $\alpha = \left(\frac{n}{\pi}\right)^{2/5}$. Then the smallest quantum

number that will give the value of the fine structure constant will be given by, n= $1.429101876 \times 10^{-5}$. This is the lower limit for the quantum theory.

Then on the quantum gravitational scale, in which the Bohr's quantized

energy is equated to our energy $\frac{2\hbar c^3}{e}\left(\frac{\pi\varepsilon_0}{G}\right)^{1/2}$, we obtain the relationship between

the principal quantum number n, the gravitational coupling constant α_G and

the fine structure or the electromagnetic constant α_E as, $n^2 = \alpha_G \alpha_E^5$ or

$n = \alpha_G^{1/2} \alpha_E^{5/2} \sim 3.493 \times 10^{-25}$. This value implies an upper bound on the energy states for a combined theory of gravity and quantum mechanics between two protons.

(5)Last but not least, we could write a modified Unruh- Davis effect as,

$T = \frac{\hbar c^3}{k_B}\left(\frac{\varepsilon_0 \alpha}{\pi G}\right)^{1/2}$, when it is equated to the Hawking temperature effect

$T = \frac{\hbar c}{k_B}(\Lambda)^{1/2}$, where Λ is the curvature of space, we obtain the curvature as,

$\Lambda = \frac{c^4 \varepsilon_0}{\pi e^2 G}\alpha$, then for $\alpha = \frac{e^2}{4\pi\varepsilon_0 \hbar c}$, we obtain the maximum curvature for quantum gravitational effects as, $\Lambda = \dfrac{c^3}{4\pi^2 G \hbar}$. But when $\alpha = \dfrac{Gm^2}{\hbar c}$, we obtain the curvature for quantum electrodynamics effects as, $\Lambda = \dfrac{m^2 c^3 \varepsilon_o}{\pi e^2 \hbar}$.

CHAPTER1:The Study Of Gravitational And Electromagnetic Radiation Intensity On a Quantum Scale as a Basis For the Development of the Theory of Quantum Gravity.

Albert Einstein was one of the physicists who attempted to develop a classical unified field theory but in vain. Other mathematicians and physicists who have attempted like Einstein to develop a unified field theory include among the many Hermann Weyl, Theodor Kaluza and R. Bach, but due to the continual development of quantum theory and the difficulties encountered in developing a quan-tum theory of gravity most of the physicists have gave up working on the unified field theories as of date.

Although most of the scientists have abandoned classical theories, still they remain the only means through which the quantum theory of gravitation can be created and thereafter be unified with the other fundamental theories in physics as this book directs.

At present, one of the deepest problems in theoretical physics is harmonizing the theory of general relativity, which describes gravitation, and applications to large-scale structures (stars, planets, galaxies), with quantum mechanics, which describes the other three fundamental forces acting on the atomic scale. However there is a connection between the fundamental forces of electromagnetism and gravitation through which the quantum theory of gravity can be created. This connection is based on two rules, 1) the fundamental significance of the finite and invariant velocity of light in inertial reference frames in the special theory, and 2) the reliance of the general theory of relativity upon the special theory of relativity locally in space-time. The connection between the fundamental forces of electromagnetism and gravitation follows immediately from these two points (Douglas M. Snyder).

To unite electromagnetism with gravity and thereafter with the quantum theory, we must create in principle a clear and analytical study of the radiation of gravitational energy by an atom.

Gravitational radiation is produced when massive bodies accelerate. This radiation is difficult to detect due to the weakness of the gravitational force.

It can only be detected under vigorous observations of the radiations from supernovae and collisions of black holes. The study of the gravitational radiation would come straight from the Bohr's theory of an atom but it proves difficult since one cannot even explain the intensity of the electromagnetic wave from the theory. Rather the intensity of radiation emitted from an atom is studied using the known formula for the intensity in electromagnetism (EB /μo). To clearly understand the intensity of the electromagnetic wave, one needs to develop a formula for the intensity of a wave on a quantum scale. Once such is formulated, it would then become easy to perform calculations for the intensity of the gravitational waves.

The unification of electromagnetism with gravity for a long time has been difficult, in a statement written by Jeroen Burgers (2009) it is clear that the two interactions can be unified into a single interaction. In his statement Jeroen writes, "If an electrical charge is accelerating, it will emit radiation, e.g, as radio waves from an antenna. Correspondingly, according to GR, accelerating masses should emit gravitational radiation, a propagating deformation of space-time".

In this study our aim is to unite the electromagnetic interactions with the gravitational interactions by developing a relationship between the intensity of the gravitational and electromagnetic waves.

To differ from Bohr's model of an hydrogen atom, it is hereby theorized, that an electron moving in an atom will possess an energy due to the electric, magnetic and gravitational forces acting on it as formulated below,

$$W_p = \frac{m_gE_e}{B_{ev}}r = \frac{F_GF_e}{F_B}r \quad (1)$$

Where $F_G = mg$ is the gravitational force, $F_e = Ee$ is the electric force on an electron in vicinity of an electric field and $F_B = Bev$ is the magnetic force

To determine the strength of the electromagnetic force on a quantum scale, we borrow an analogy from the theory of quantum mechanics by which the quantized angular momentum is deduced from the fine structure constant and denoting the coupling constant to behave as the principle quantum number (Remember our aim here is to deduce the intensity of a wave emitted from an atom due to an electron performing Bohr's orbits), In formula we express the angular momentum as,

$$\frac{Ke^2}{c} = n\hbar \quad _2$$

Where $\hbar$ is the reduced plank constant, c is the speed of light, k is the coulomb constant ($k = 1/4\pi\varepsilon_0$, ε_0 is permeability of free space) and n is the principle quantum number.

Since the gravitational force is almost negligible in an atom, it becomes a catastrophe to treat it as a quantum mechanical effect. In quantum mechanics the angular momentum of an electron is quantized in units of $n\hbar$ while in gravitational mechanics, there is no such thing as quantization, which is why we treat gravitation classically.

Then the formula for the angular momentum due to the gravitational force will be given by,

$$\frac{Gm^2}{c} = mvr \quad _3$$

It is therefore evident that the gravitational descriptions of an electron can only be treated classically. This is why it has proven difficult to merge gravity with quantum mechanics. However such a problem has been solved here, by considering the assumptions below,

The speed of light in both quantum and gravitational processes is a constant and therefore if we substitute the speed of light from eqn (2) into eqn(3) we get the expression for the angular momentum as,

$$mvr = \frac{F_G}{F_e} n\hbar \quad _4$$

We have thus introduced the ratio of the gravitational force to the electric force in the formula for Bohr's quantized angular momentum. This ratio represents the negligible gravitational effects in an atom. It is therefore a correction to the Bohr's atomic model. It is a small number like the principle quantum number that is almost negligible in the calculation of the radiations emitted from an atom.

Because the gravitational force can be expressed in many ways by using Eqn(1), we can deduce the power carried by the electromagnetic wave due to the motion of an electron in an atom. Making F_G the subject from equation (1) and substituting for it in equation (4), we get the power as,

$$F_{BC} = \frac{2\pi r^2 \lambda m v F_e^2}{nh^2}$$

Since the de Brogile wave length is $\lambda = h/mv$ and the surface area of the sphere $A = 4\pi r^2$. Then the Intensity of a wave from a particle exhibiting both wave and particle properties is

$$\frac{F_{BC}}{A} = \frac{F_e^2}{2nh} = \frac{E^2 e^2}{2nh\,6}$$

Keeping other factors constant we have theorized that, the intensity of a wave is proportional to the square of the electric field, a fact that would be impossible to deduce in Bohr's atomic model. The above formula can only be deduced if only we take into account (in theory), the combined effects of gravity and electromagnetism. On the other hand, *If we let the power of the electromagnetic wave be $P = F_{BC}$, and n be the fine structure constant $\alpha = ke^2/\hbar c$, then the equation for the intensity of the electromagnetic wave comes out clearly as,*

$P = EB/\mu o = 2\varepsilon_0 E^2 c$, *Where μo is the permeability of free space.*

Gravitational waves are harder to generate than the electromagnetic waves, simply because, due to the conservation of the angular momentum, there's no dipole gravitational radiation meaning that, the dominant mode of the gravitational radiation is quadrupole radiation (see Einstein quadrupole formula). Although the gravitational waves are hard to generate, there is a possibility of studying their character and property by calculating their intensity, a fact that has proved to be difficult since there is no formula in quantum gravitational theories that can prove it.

The intensity of the gravitational wave can be deduced if only we assume that, the electric force in quantum mechanics can only be expressed as, $(Ee = \frac{n^2 \hbar c}{8\pi r^2})$. Where r is the radius of orbit of an electron in an atom, when this radius approaches the schwarzichild's radius $(r = \frac{Gm}{c^2})$ for Black holes, then the electric force is expressed as, $(Ee = \frac{n^2 \hbar c^5}{8\pi G^2 m^2})$. But the intensity in quantum mechanics was deduced in Eqn6. Therefore the intensity of the gravitational wave will be given by,

$$\frac{F_{BC}}{A} = \frac{E^2 e^2}{2nh} = \frac{n^3 \hbar c^{10}}{256\pi^3 G^4 m^4 7}$$

At the surface of the sun the intensity of the solar radiation is about $6.33 \times 10^7 W/m^2$, which means that, we will require a mass of about $8.8965 \times 10^{19} kg$, to obtain that intensity from our formula above.

For the sun of mass $1.989 \times 10^{30} kg$, the intensity is so small with a value of only, $0.253 \times 10^{-33} W/m^2$ due to the weakness of gravity.

At the Planck mass $M_P = \sqrt{\dfrac{\hbar c}{G}}$, the intensity will be given by, $I = \dfrac{c^8}{256\pi^3 G^2 \hbar}$.

CHAPTER 2: The Magnetic Fields at Which the Theories of Quantum Electrodynamics and Quantum Gravity Become Analogous (Field limit)

The scale in quantum electrodynamics (QED), above which the electromagnetic field is expected to become non linear, also called the Schwinger limit, was first derived by Fritz Sauter in 1931. However In this section we develop a mechanism (which differs from Fritz's approach) through which the Schwinger limit is deduced using a dimensionless number, which gives the critical magnetic field in quantum electrodynamics when its value is equal to the electromagnetic coupling constant and in the same way gives the critical magnetic field in Quantum gravity when its value is equal to the gravitational coupling constant.

According to D.A. Leahy, the application of quantum electrodynamics in strong magnetic fields only fairly recently has been a subject of interest. The motivation for this study was the discovery of Neutron stars with very high magnetic fields of orders 10^{12} -10^{13}G.

With the discovery of magnetars, quantum electrodynamics calculations which are valid for very high fields became of great interest. The critical value of the magnetic field is defined as $B = \frac{m^2 c^2}{\hbar e} = 4.414 \times 10^{13}$G.

However, there is a value of the magnetic field that is bigger and stronger than the critical magnetic field strength in Quantum electrodynamics and this magnetic field is of orders of magnitude 10^{52}G. Such a big value has not been deduced in any existing scientific literature and that is the reason why I take pleasure in deriving it here and hence call it the "quantum gravity threshold".

From equation (6) $\frac{F_{BC}}{A} = \frac{F_e^2}{2nh}$, if we let the magnetic force to be equal in magnitude and strength to the electric force, we create two relationships, 1) the force on a particle falls off as the area it occupies and 2) the force falls off as the principle quantum number.

$$\text{Force (F)} = \frac{2hc}{A} = \frac{Bev}{n}$$

If n was the fine structure constant ($ke^2/\hbar c$, $k = 1/4\pi\varepsilon_0$), the speed of light in vacuum being $c = \lambda f = \lambda\omega/2\pi$ and the velocity of a particle in the magnetic field is $v = \omega r$ where ω is the angular frequency for circular motion we have

$$\frac{F_c}{F} = \frac{em}{2B\lambda\hbar\varepsilon_0 8}$$

Where $F_c = m\omega^2 r$ is the centripetal force

We have thus derived a general formula for the coupling of forces. Then the *Schwinger limit* in quantum electrodynamics for the critical magnetic field can be deduced from the above expression when we set the ratio of the forces to be equal to the electromagnetic coupling or fine structure constant as, $\dfrac{F_c}{F} = \dfrac{ke^2}{\hbar c}$.

$$B_{QED} = \frac{2\pi mc}{\lambda e}$$

For a particle with deBrogile wavelength $\dfrac{2\pi\hbar}{mc} = \lambda$, the quantum electrodynamics threshold is given by,

$$B_{QED} = \frac{m^2 c^2}{\hbar e} = 4.3697 \times 10^{13} G_9$$

However, for $\dfrac{F_c}{F} = \dfrac{Gm^2}{\hbar c}$, the gravitational coupling constant, and $\dfrac{2\pi\hbar}{mc} = \lambda$, the deBrogile wavelength, the *quantum gravity threshold* is given by a value,

$$B_{QG} = \frac{ec^2}{4\pi G\hbar\varepsilon_0} = 1.8423 \times 10^{52} G_{10}$$

We have thus deduced the constant magnetic field carried by an electron in the combined quantum electromagnetic and gravitational fields. The fact that the formula has the fundamental constant of electricity (ε_0), relativistic quantum mechanics (c, $\hbar$) and Gravity (G), is an indication that this is the quantum gravity limit or a scale at which the electromagnetic field is expected to become non linear.

CHAPTER3:Derivation of the temperature and entropy of black holes

"if the semi-diameter of a sphere of the same density as the sun were to exceed that of the sun in the proportion of 500 to 1, a body falling from an infinite height towards it would have acquired at it's surface greater velocity than that of light, and consequently supposing light to be attracted by the same force in proportion to its vis inertiae, with other bodies, all light emitted from such a body would be made to return towards it by its own proper gravity."

-JohMichell

The development of general relativity followed a publication of acceleration under special relativity in 1907 by Albert Einstein. In his article, he argued that any mass will "Distort" the region of space around it so that all freely moving objects will follow the same curved paths curving toward the mass producing the distortions. Then in 1916, Schwarzschild found a solution to the Einstein field equations, laying the groundwork for the description of gravitational collapse and, eventually, black holes.

By definition, a black hole is an astronomical object with a very strong gravitational effect, which disturbs particles across its event horizon. It is also true from the theory of general relativity, that even light can not escape its gravitational pull. These objects have puzzled the minds of great thinkers for many years. History puts it that, they were first predicated by John Michell and Pierre-Simon Laplace in the 18th century but David Finkelstein was the first person to publish a promising explanation of them in 1958 based on Karl Schwarz child's formulations of a solution to general relativity that characterized black holes in 1916.

In 1971, Hawking developed what is known as the second law of black hole mechanics in which the total area of the event horizons of any collection of classical black holes can never decrease, even if they collide and merge. This is similar to the second law of thermodynamics which states that, the entropy of a system can never decrease. In 1972 Bekenstein proposed an analogy between black hole physics and thermodynamics in which he derived a relation between the entropy of black hole entropy and the area of its event horizon.

In 1974, Hawking predicted an entirely astonishing phenomenon about black holes, in which he ascertained with accuracy that black holes do radiate or emit particles in a perfect black body spectrum. Hawking was able to produce in result the temperature of a black hole and proposed that, this temperature was proportional to the surface gravity of a black hole.

Both the temperature and entropy of a black hole have been deduced in literature using different approaches which have proved to be true but only lack one ingredient and that is, the description of a force like the Newtonian force of gravity, which acts to pull matter from the event horizon up to the point of its singularity. If this force was true it would be able under general conditions to derive the Reissner- Nordstrom metric for the charged non-rotating black hole as we are to see

Temperature of a black hole

It is here by hypothesized that, the gravitational field will create particles and emit them only if the electromagnetic force of such particles were equal to the

$$F = \frac{Me}{r}\sqrt{\frac{Gp}{2\hbar\varepsilon_0\lambda}}$$

force (unknown in literature) .Where p, is the momentum of a particle. under general conditions, the force given will reduce to the Reissner-Nordstrom metric as given here, if the momentum of an electron at a distance r from the singularity point to the event horizon is related to the de Brogile wavelength as $p = \frac{2\pi\hbar}{\lambda}$, and both the distance r and wavelength λ was the product of the speed of light c and the period T as r=cT and $\lambda = cT$, then the force will be given by $F = \frac{Mp}{r\hbar}\sqrt{\frac{Ge^2}{4\pi\varepsilon_0}}$, but since $\frac{p}{2\pi\hbar} = \frac{1}{\lambda}$, then we have,

$$F = \frac{2\pi M}{T^2}\sqrt{\frac{Ge^2}{4\pi\varepsilon_0 c^4}}$$, this reduces to $F = \frac{2\pi M}{T^2}r_q$, where $r_q = \sqrt{\frac{Ge^2}{4\pi\varepsilon_0 c^4}}$ is the Reissner-Nordstrom radius of a charged black hole.

Having derived the Reissner-Nordstrom metric from our force formula, we now return to our exercise of deriving the temperature of a black hole. We consider a particle with charge e, exhibiting deBrogile wave properties of momentum and wavelength from the centre of mass M of a black hole. We then assume that this particle experiences an electromagnetic force due to the magnetic and electric field created by other particles in its surrounding area. The same particle also experiences a force due to the strong gravitational field emanating from the black hole. Equating the two forces as

$$\frac{Me}{r}\sqrt{\frac{Gp}{2\hbar\varepsilon_0\lambda}} = \frac{e^2}{4\pi\varepsilon_0 r^2}$$, from this expression we obtain the momentum of a

particle as $$p = \frac{\hbar e^2 \lambda}{2\pi A\varepsilon_0 GM^2}$$. This is the momentum possessed by a particle (emitted by the gravitational field of a black hole) at the surface of the event

horizon, where $A = 4\pi r^2$ is the spherical surface area of the horizon.

For relativistic effects, the kinetic energy of a particle will be related to its momentum by K.E=pc and to the Boltzmann's law by K.E=kT, where k is the Boltzmann's constant and T is the absolute temperature. By similarity we can equate the two energies as pc=kT, then from the equation of momentum we can obtain the temperature as,

$$T = \frac{\hbar e^2 \lambda c}{2\pi A\varepsilon_0 GM^2 k}.$$

Expressing the permittivity of free space in terms of the permeability of free

space $$\varepsilon_o = \frac{1}{\mu_0 c^2}$$, we obtain the Hawking temperature of a black hole as,

$$T = \left(\frac{4e^2\mu_0\lambda}{AM}\right)\frac{\hbar c^3}{8\pi GMk}$$

In a more general form, in terms of energies it can be expressed as,

$$T = \left(\frac{4e^2\lambda}{A\varepsilon_0 Mc^2}\right)\frac{\hbar c^3}{8\pi GMk}\,11$$

Entropy of a black hole

In an attempt to prevent the violation of the generalized second law of thermodynamics, Bekenstein proposed a universal upper bound on the ratio entropy to energy for bounded systems (Phys RevD23, 287-1981), which was later rejected by Unruh and Wald in 1982. They proposed a thought experiment in which a box lowered down into a black hole felt an effective buoyancy force which was caused by the acceleration radiation felt by the box near the black hole. They argued further that, this buoyancy force would guarantee a lower bound on the energy gain of the black hole, hence saving the generalized second law without a need for entropy bound.

In this section we give a formula for the buoyancy force which is different from the Unruh and Wald formula which appeared in their 1982 paper.

At a distance r from the center of mass m of a black hole, the buoyancy force is given by,

$$F_B = \frac{rc^6}{8G^2m} \quad 12$$

From the above force formula the energy gain by the black hole will be given by,

$$W_B = \frac{Ac^6}{32\pi G^2 m}$$

Where, A is the area of the event horizon. Since entropy is the ratio of energy to temperature, $S_B = W_B / T_B$ and temperature of a black hole is known from equation 11, then the entropy of a black hole is given by,

$$S_B = \frac{Akc^3}{4G\hbar}\left(\frac{A\varepsilon_0 Mc^2}{4e^2\lambda}\right) \quad 13$$

The Earliest Period of Time[1] in the History of the Universe[2]

Observations have suggested that the universe began 13.7billion years ago. The universe was so hot with particles having a very high energy, in its earlier phase. The evolution then proceeded with this energy forming the first protons, electrons and neutrons, then nuclei and finally atoms. The microwave background was also emitted during the formation of the neutral hydrogen. Finally the structure of the universe was formed when matters aggregated into the first stars and quasars and on large scale clusters of galaxies and super clusters were formed.

In cosmology[3], the Planck epoch , named after Max Planck[4], is the earliest period of time[5] in the history of the universe[6], from zero to approximately 10^{-43} seconds, it is at this time that quantum effects[7] of gravity[8] were significant. At this

1. http://en.wikipedia.org/wiki/Time

2. http://en.wikipedia.org/wiki/Universe

3. http://en.wikipedia.org/wiki/Physical_cosmology

4. http://en.wikipedia.org/wiki/Max_Planck

5. http://en.wikipedia.org/wiki/Time

6. http://en.wikipedia.org/wiki/Universe

7. http://en.wikipedia.org/wiki/Quantum_mechanics

period approximately 1.37×10^{10} years ago all fundamental forces[9] were unified. The state of the universe during the Planck epoch was unstable, tending to evolve and giving rise to the familiar manifestations of the fundamental forces through a process known as symmetry breaking[10]. It is currently believed that the Planck epoch inaugurated the Grand unification epoch[11], and that symmetry breaking quickly led to the era of cosmic inflation[12], the Inflationary epoch[13], during which the during which the universe greatly expanded in scale over a very short period of time (see Wikipedia Planck epoch)

The first seconds of the universe can be calculated if we apply the law of attraction to the box lowered into a black hole, we assume that there is a force acting in opposite direction to the buoyancy force (Eqn12). The forces acting on the box are different in a way that, the force expressed in equation 12 is proportional to the schwarzichild's radius while the force acting in the opposite direction is proportional to time measured in seconds. Therefore we have two different forces, one taking a particle through a distance and the other through timelines.

Therefore in time t for any particle moving through space, the force acting on such a particle will be given by,

$$F_t = \frac{e^2 c^5 \hbar}{32 \pi \varepsilon_0 G^3 m^4} = t \frac{c^7}{16 \pi G^2 m}$$

Where $t = \frac{e^2 \hbar}{2 \varepsilon_0 c^2 G m^3} = 2.54821 \times 10^{-68} \frac{1}{M^3} \ (s)$

Thus keeping other factors constant, the time is inversely proportional to the cube of the mass of a particle. For the Planck mass of $2.1765 \times 10^{-8} kg$, the earliest period of time in the universe is $2.472 \times 10^{-45} s$. Then the total mass responsible for the current age of the universe (13.82 billion years) is, $3.880 \times 10^{-29} kg$

---

8.   http://en.wikipedia.org/wiki/Gravity

9.   http://en.wikipedia.org/wiki/Fundamental_force

10.   http://en.wikipedia.org/wiki/Symmetry_breaking

11.   http://en.wikipedia.org/wiki/Grand_unification_epoch

12.   http://en.wikipedia.org/wiki/Cosmic_inflation

13.   http://en.wikipedia.org/wiki/Inflationary_epoch

Black hole radiations/ Hawking radiations

From the book: THE TUTOR'S REFERENCE by Balungi Francis

By definition as of Wikipedia, a black hole is a mathematically defined region of space time exhibiting such a strong gravitational pull that no particle or electromagnetic radiation can escape from it. Many theories have been created to explain the properties of the black hole but the theory created here is far more different from the other theories although it may give the same results. Using a quite different approach towards solving a problem is efficient since it comes with it new predictions in the process which could have been hidden in other approaches. Below we try to present adhoc proofs-laws that may be of help in building our theory about black holes, note; these proofs can be derived mathematically from equations 1 up to 4 above but for purposes of simplicity they have been listed here below, however their derivations will be given in the last chapters of this book.

The laws or equations:

It is well known that the electric field is force per unit charge but here a generalized equation for an electric field created by an electron exhibiting wave properties in the nucleus of an atom in the gravitational field on a quantum scale is given by

$$E = \frac{1}{r}\sqrt{\frac{Gm^3 f}{2\hbar \varepsilon_0}}$$

.................(5)

Then the electric force in this case will be formulated as

$$F_1 = \frac{e}{r}\sqrt{\frac{Gm^3 f}{2\hbar \varepsilon_0}}$$

........................(6)

The surface area at a radius r of orbit of an electron of mass m around the nucleus of an atom in a wave like manner is given by

$$\text{surface area}(A) = \frac{\lambda \mu_0 e^2}{m}$$

..............(7)

The time taken by the magnetic field B of an electron to pass through a given surface is

$$\text{time}(t) = \frac{\lambda \varepsilon_0 AB}{e} \cdots$$

........(8)

Note: the above expression is the same as Faraday's induction law.

The gravitational force acting on all matter in the universe or the modified gravitational force is given as

$$F_2 = \left(\frac{Gm^3}{r^2}\right)\left(\frac{e}{2B\lambda\hbar\varepsilon_0}\right) \quad \text{............(9)}$$

The above formulas are important in deriving the formula for the temperature, entropy and the time taken by a black hole to evaporate as shown below;

Temperature of a black hole

It is known that the kinetic energy KE of molecules in the Boltzmann hypothesis is related to the temperature of the body in question in this case a black hole (in relation to the black body) by $KE = \varphi T$ where φ is Boltzmann's constant. The formula for the kinetic energy can be derived by using a hypothesis that the electromagnetic force – coulombs force is equal to eqn6 as

$$\frac{ke^2}{r^2} = \frac{e}{r}\sqrt{\frac{Gm^3 f}{2\hbar\varepsilon_0}}$$

On squaring both sides of the equation, cancelling like terms and taking into account that the frequency of an electron is $f = \frac{v}{\lambda}$, then the kinetic energy of an electron inside the black hole is given by

$$KE = \frac{\lambda\mu_0 e^2}{A}\frac{c^3\hbar}{8\pi Gm^2}$$

Since the surface area is given as from eqaution7 then the kinetic energy of molecules or particles (for an ideal gas) within the black hole will be given by

$$KE = \frac{c^3\hbar}{8\pi Gm} = T\varphi \quad (10)$$

Then from Boltzmann's relationship the temperature of the black hole is formulated as

$$T = \frac{c^3\hbar}{8\pi Gm\varphi}(11)$$

The entropy of the black hole

By definition entropy is a measure of disorder. To solve the entropy of black holes we shall consider a very complex argument about the entropy in question. We assume that the modified gravitational force given by equation 9 is identical to the modified electric field given by equation6 as,

$$\left(\frac{Gm^3}{r^2}\right)\left(\frac{e}{2B\lambda\hbar\varepsilon_0}\right) \equiv \frac{e}{r}\sqrt{\frac{Gm^3f}{2\hbar\varepsilon_0}}$$ in otherwise the two forces are equal but opposite. Then squaring both sides of the equation and multiplying through by Gc^5 one obtains a new relation of forces on both sides given as

$$\frac{tc^7}{16\pi G^2 m} = \frac{Ac^6}{32\pi rm G^2}$$

Both the left and right hand side represent a force. From the left hand side t is the expression of time given by $$t = \frac{\hbar e^2}{2m^3 c^2 G\varepsilon_0}.$$ Note: the left hand side force is the pull of matter inside the black hole while the right hand side force is the force acting on particles or matter at the surface of the black hole.

Since the heat is the product of the force on a particle and the distance r from the centre of the black hole, then using the force on the right hand side of the above equation the heat will be given by

$$Q = \frac{Ac^6}{32\pi mG^2}$$

Remember the temperature of the black hole is also known from equation9 and by definition the entropy of the system is the change in heat per unit temperature $\frac{Q}{T}$, then the entropy of the black hole will be given by

$$S = \frac{A\varphi c^3}{4G\hbar} \quad (12)$$

This implies that the entropy of a black hole is proportional to its surface area.

The time taken by a black hole to evaporate

Assuming that particles that formed a black hole are moving away or are separating from it after a given time of its existence, if we measure the relative speed of these particles in relation to the energy they carry we obtain a relation ship given by

$$\frac{v^2}{c^2} = \frac{8\pi G}{c^2}\left(\frac{W}{8\pi r}\right) \quad (13)$$

Where v is the velocity of these particles as measured relative to the speed of light c and W is the energy carried by the particles as they move away from the centre of the black hole at a distance r.

If we let the force causing the particles to separate from the black hole be given as $\dfrac{Gm^3 e}{2r_\lambda B\hbar\varepsilon_0}\dfrac{v}{c}$, then the energy of these particles will be given by

$$W = \frac{Gm^3 e}{2r_\lambda B\hbar\varepsilon_0}\frac{v}{c}$$

Substituting this in equation11, we obtain a relation ship of time as given by the law 3 of equation 8 as

$$t = \frac{v^2}{c^2}\left(\frac{\pi G^2 m^3}{\hbar c^4}\right)$$

The velocity of the particles in the astronomical lab will be measured as v= 4.193E6 m/s and since the speed of light is a constant then the time taken by a black hole to evaporate is given by

$$t = \frac{5120\pi G^2 m^3}{\hbar c^4} \quad (14)$$

CHAPTER4: The Art of Reductionism

Scientific reductionism is the idea of reducing complex interactions and entities to the sum of their constituent parts, in order to make them easier to study (explorer.com). It is based on the idea that science can be used to explain everything by a mere look at the individual constituent processes.

There are three types of reductionism, that is, ontological, methodological and theory reduction. In this section we shall emphasize theory reduction because we have a great deal of reducing known laws of physics from a somewhat simple rule. This was the case when Kepler's laws of the motion of planets and Galileo's theories of motion were reduced to the Newtonian theories of mechanics.

Newtonian Mechanics became a more general theory simply because all the explanatory power of Kepler's and Galileo's laws was contained in it. Therefore theoretical reduction is considered as the reduction of one explanation or theory to another.

The most interesting thing about this section is that, during the process of reduction we create a relationship between the known law to another law explaining the same thing but unknown to the entire physics community. **For example in the reduction of the Weidman Franz- Lorenz law we create in a process a law for the thermal conductivity of gravito-electric phenomenon.**

Therefore reductionism is deriving something complicated from something simple. For example in the derivation of the Weidman Franz law we set a formula that states that, the electric force (Ee) on an electron is proportional to the gravitational force at the schwarzichild's radius $\left(\dfrac{c^4}{8\pi G}\right)$ but inversely proportional to the gravitational coupling constant $\left(\dfrac{Gm^2}{\hbar c}\right)$ as given below,

$$F = \frac{n^2}{\alpha_g} f_g$$

Where n, is the principle quantum number.

On squaring the above equation we obtain the square of the electric field as,

$$E^2 = \frac{n^4 c^4}{G^2 e^2 m^2}\left(\frac{c^3 \hbar}{8\pi Gm}\right)^2$$

From the formular for the temperature of the black hole, the function $\overline{\frac{c^3 \hbar}{8\pi Gm}}$ is related to temperature as kT, and then the law for thermal conductivity will be reduced as,

$$\frac{\pi^2 E^2 G^2 m^2}{3Tc^4} = \left(\frac{n^4 \pi^2}{3}\right)\left(\frac{k}{e}\right)^2 T$$

The left hand side represents the ratio of the thermal conductivity K to the electric conductivity δ, which is the Weidman Franz law. From the above reduction we have generated an important rule given by,

$$\frac{K}{\delta} = \frac{1}{3}\left(\frac{\pi Gm}{c^2}\right)^2 \frac{E^2}{T} = \frac{\pi^2 r_s^2}{3}\frac{E^2}{T}$$

The above formula explains the thermal properties of black holes at the schwarzichild's radius r_s

CHAPTER5: On the Bohr's model of the Hydrogen atom from a Quantum Gravity perspective and the deduction of the earliest period of time

In the early 20th century[1], Ernest Rutherford[2] experiments established that atoms[3] consisted of a diffuse cloud of negatively charged electrons[4] surrounding a small, dense, positively charged nucleus. Given his experimental data, it was quite natural for Rutherford to consider a planetary model for the atom, the Rutherford model[5] of 1911, with electrons orbiting a sun-like nucleus. This model was a difficulty. The laws of classical mechanics predict that the electron will release electromagnetic radiation[6] as it orbits a nucleus. Because the electron would be losing energy, it would gradually spiral inwards and collapse into the nucleus. This was a disaster, because it predicted that all matter was unstable.

To overcome this difficulty, Niels Bohr[7] proposed, in 1913[8], what is now called the Bohr model of the H atom. The model's key success laid in explaining the Rydberg formula[9] for the spectral emission lines[10] of atomic hydrogen. Not only did the Bohr model explain the reason for the structure of the Rydberg formula, but it provided a justification for its empirical results in terms of fundamental physical constants.

This section looks at the model in a very different way than that of Bohr. The fact that all accelerated particles do emit electromagnetic radiations is taken into account and therefore the acceptance for the unstableness of all matter is

---

1. http://en.wikipedia.org/wiki/20th_century

2. http://en.wikipedia.org/wiki/Ernest_Rutherford

3. http://en.wikipedia.org/wiki/Atom

4. http://en.wikipedia.org/wiki/Electron

5. http://en.wikipedia.org/wiki/Rutherford_model

6. http://en.wikipedia.org/wiki/Electromagnetic_radiation

7. http://en.wikipedia.org/wiki/Niels_Bohr

8. http://en.wikipedia.org/wiki/1913

9. http://en.wikipedia.org/wiki/Rydberg_formula

10. http://en.wikipedia.org/wiki/Emission_line

considered in due respect. In fact Bohr's ideas never required classical mechanics simply because it could not conform to the experimental observations of the spectrum of the Hydrogen atom that were obtained by Rydberg using his formula.

To merge gravity with Planck's quantum theory by then was also a problem at hand and therefore Bohr had to forego the problem by introducing in his theory adhoc postulates, and this could have been the reason why Einstein found problems in merging gravity with electromagnetism in what is called "The Grand unified field theory", of which he had to question the problem with the quantum theory and therefore request for a complete quantum theory. From Bohr's model many theories have been formed each building from the ideas of the model, but a certain point is reached where the theories can not conform well to the known laws of nature and therefore regarded as failures, which of course in their judgments is true. The problem is seen to come from exactly the roots of quantum mechanics.

The aim of this section is therefore to produce a generalized theory of atomic structure that incorporates in it gravity and quantum mechanics and thus predict the properties of the universe at the Planck era.

Methodology

The Hydrogen atom exists in certain stationary states of discrete energies. The acceleration due to gravity of an electron in orbit around the nucleus will cause the atom to emit radiations (radiate energy) and thus make the atom unstable. The acceleration (g) falls off with time t provided the radius of orbit of the electron R is a constant thus the acceleration due to gravity is given by;

$g = R/\Delta t^2$ (1)

The rate of change of energy P radiated as a result of the above acceleration will depend on the constants c (speed of light) and G (universal gravitational constant), hence;

$P = c^5/G$ (2)

The power and time must be re- quantized in units of $\hbar = h/2\pi$ where h is Planck constant, hence

$P\Delta t^2 = n^2\hbar$ (3)

Where n= 1,2,3........ is the principle quantum number.

But the total energy of the atom in the various energy states is $W = -ke^2/R$ where k is the Coulomb constant and e is the elementary charge. Since Δt^2 is known from Eqn1 and P from Eqn2 then using Eqn3 the radius is given by

$R = n^2 Gg\hbar / c^5$ (4)

From which the total energy is given by,

$W = -ke^2 c^5 / n^2 Gg\hbar$ (5)

From the Bohr-Einstein frequency (f) condition, applied to a transition from a level with $n = n_i$ to a level with $n = n_f$, The energy of a photon emitted by a hydrogen atom is given by the difference of two hydrogen energy levels

$hf = E_i - E_f$

Finally we get since frequency $f = c/\lambda$, where λ is the wavelength

$1/\lambda = [ke^2 c^4 / 2\pi G\hbar^2][1/g][1/ n_f^2 - 1/ n_i^2]$ (6)

The equation obtained above shows some how a great significance of gravity in the quantum theory. So far it states that regardless of the levels in the transitions of an atom the acceleration due to gravity of the particles in the atom do greatly affect the nature of its spectrum.

Results

The quantity $[ke^2 c^4 / 2\pi G\hbar^2]$ is the inverse of the square of time t and therefore

$1/t^2 = [ke^2 c^4 / 2\pi G\hbar^2]$, from which the time is obtained as $t = 1.58873 \times 10^{-42}$ s. This is the earliest period of time in the history of the universe.

Comparing Eq6 with Bohr's model, here we shall equate the Rydberg constant $[k^2 e^4 m/4\pi c\hbar^3]$, where m is the mass of the particle, to the constant $[ke^2 c^4 / 2\pi G\hbar^2][1/g]$. Doing this generates an acceleration given by $g_a = [$

$$m = \sqrt{\left(\frac{\hbar c}{G}\right)}$$

$2\hbar c^5/ke^2 Gm]$, then at the Planck epoch when _____ the acceleration

reduces to $g = \dfrac{8\pi\varepsilon_0}{e^2}\sqrt{\left(\dfrac{\hbar c^7}{G}\right)}$. Then At the Schwarz child's radius $R = Gm/c^2$ the acceleration is $g_b = c^4/Gm$ which gives an equation for the spectrum as 1/

$\lambda = [/ 2\pi \, a_o][1/ n_f{}^2 - 1/ n_i{}^2]$ where a_o is the first Bohr radius $[\hbar^2/ mke^2] = 5.28 \times 10^{-11}$m.

The interesting part of it is that the ratio $g_b / g_a = [ke^2/2\hbar c]$ is the fine structure constant.

CHAPTER6: Deduction of the maximal magnetic field, radiation intensity, quantum hall effect and the laws of black hole mechanics from a proposed theory of quantum gravity

Quantum gravity is the field of theoretical physics[1] that tries to unify quantum mechanics[2] with general relativity. Quantum mechanics describes the three fundamental forces of nature[3] while general relativity[4] is a theory of the fourth fundamental force: gravity[5]. The goal every one is waiting for to emerge from this unification is a "theory of everything[6]", or "Grand Unified Theory[7]" (GUT). So many researches have been conducted in line with the theory, for example in 1986[8], Abhay Ashtekar[9] reformulated Einstein's field equations of general relativity using what have come to be known as Ashtekar variables[10], a particular flavor of Einstein-Cartan theory[11] with a complex connection. He was able to quantize gravity using gauge field theory[12]. In the Ashtekar formulation, the fundamental objects are a rule for parallel transport[13] and a coordinate frame known as a vierbein[14] at each point. Because the Ashtekar formulation was background-independent, it was possible to use Wilson loops[15] as the basis for

1. http://en.wikipedia.org/wiki/Theoretical_physics

2. http://en.wikipedia.org/wiki/Quantum_mechanics

3. http://en.wikipedia.org/wiki/Fundamental_interaction

4. http://en.wikipedia.org/wiki/General_relativity

5. http://en.wikipedia.org/wiki/Gravitation

6. http://en.wikipedia.org/wiki/Theory_of_everything

7. http://en.wikipedia.org/wiki/Grand_Unified_Theory

8. http://en.wikipedia.org/wiki/1986

9. http://en.wikipedia.org/wiki/Abhay_Ashtekar

10. http://en.wikipedia.org/wiki/Ashtekar_variables

11. http://en.wikipedia.org/wiki/Einstein-Cartan_theory

12. http://en.wikipedia.org/wiki/Gauge_field_theory

13. http://en.wikipedia.org/wiki/Parallel_transport

14. http://en.wikipedia.org/wiki/Vierbein

15. http://en.wikipedia.org/wiki/Wilson_loop

a nonperturbative quantization of gravity. Explicit (spatial) diffeomorphism invariance of the vacuum state[16] plays an essential role in the regularization of the Wilson loop states.

Around 1990[17], Carlo Rovelli[18] and Lee Smolin[19] obtained an explicit basis of states of quantum geometry, which turned out to be labelled by Penrose's spin networks[20]. In this context, spin networks arose as a generalization of Wilson loops necessary to deal with mutually intersecting loops. Mathematically, spin networks are related to group representation theory and can be used to construct knot invariants such as the Jones polynomial.

Quantum field theory[21] depends on particle fields embedded in the flat space-time of special relativity[22]. General relativity[23] models gravity as a curvature within space-time[24] that changes as a gravitational mass moves.

Historically, the most obvious way of combining the two (such as treating gravity as simply another particle field) ran quickly into what is known as the renormalization[25] problem. In the old-fashioned understanding of renormalization, gravity particles would attract each other and adding together all of the interactions results in many infinite values which cannot easily be cancelled out mathematically to yield sensible, finite results. This is in contrast with quantum electrodynamics[26] where, while the series still don't converge, the interactions sometimes evaluate to infinite results, but those are few enough in number to be removable via renormalization. Being closely related to topological quantum field theory[27] and group representation[28] theory, LQG is mostly

16. http://en.wikipedia.org/wiki/Vacuum_state

17.   http://en.wikipedia.org/wiki/1990

18.   http://en.wikipedia.org/wiki/Carlo_Rovelli

19.   http://en.wikipedia.org/wiki/Lee_Smolin

20.   http://en.wikipedia.org/wiki/Spin_network

21.   http://en.wikipedia.org/wiki/Quantum_field_theory

22.   http://en.wikipedia.org/wiki/Special_relativity

23.   http://en.wikipedia.org/wiki/General_relativity

24.   http://en.wikipedia.org/wiki/Spacetime

25.   http://en.wikipedia.org/wiki/Renormalization

26.   http://en.wikipedia.org/wiki/Quantum_electrodynamics

27.   http://en.wikipedia.org/wiki/Topological_quantum_field_theory

28.   http://en.wikipedia.org/wiki/Group_representation

established at the level of rigour of mathematical physics[29]. While confirming that quantum mechanics and gravity are indeed consistent at reasonable energies, this way of thinking makes clear that near or above the fundamental cutoff of our effective quantum theory of gravity a new model of nature will be needed. That is, in the modern way of thinking, the problem of combining quantum mechanics and gravity becomes an issue only at very high energies, and may well require a totally new kind of model.

The need for the paper is to understand those problems involving the combination of very large mass or energy and very small dimensions of space, such as the behavior of black holes, and the origin of the universe.

Materials and methods

The formula for the quantization of quantum gravity

The model is based on separating the gravitational field into the sum of two components; that is the background and the quantum field. The background left is one for all our calculations. But because loop gravity ignores the back ground space as a lost entity that does not occur in space, there fore the need to reconstruct quantum field theory from scratch without a background space is taken into account. I therefore suggest that the calculation should be performed by summing all possible space-times.

Quantum field theory[30] depends on particle fields embedded in the flat space-time of special relativity[31]. General relativity[32] models gravity as a curvature within space-time[33] that changes as a gravitational mass (m) moves. Assuming a spherical symmetric object that space time is of dimensions increasing from 1, 2, 3, 4...N, where N is the nth term of the dimensions. To quantize space and time is to create a space in which all of physics is quantized. The nature of the curved space surface is described by increasing powers in the Schwarzschild radius $R_s = Gm/c^2$, Hence describing the dimensions of space. Quantum mechanics explains the existence of discrete energy states in an atom, in away that the angular momentum of the atom must be quantized, which is also the case for quantum

29.   http://en.wikipedia.org/wiki/Mathematical_physics

30.   http://en.wikipedia.org/wiki/Quantum_field_theory

31.   http://en.wikipedia.org/wiki/Special_relativity

32.   http://en.wikipedia.org/wiki/General_relativity

33.   http://en.wikipedia.org/wiki/Spacetime

gravity. The equation for the quantization of the loop quantum gravity can then be written as,

$$\eta R_s + \beta R_s^2 + \mu R_s^4 + \ldots\ldots\ldots + \delta R_s^N = n\hbar \;[1]$$

Where $\eta = \sqrt{Beh}$, is the momentum of a particle probing another form of quantum mechanics, $\hbar = h/2\pi$, where h is Planck constant, $\beta = 8\pi Be$, e is the elementary charge, B is the magnetic field and finally $\mu = 256\pi^3 P/c^2$, where P is the intensity and c is the constant speed of light.

The energy equation

What changes is the form of the equation the rest remaining constant. The principle behind this is that eqn1 can be changed to any form simply for purposes of calculating complex phenomenon. The energy to which we are concerned here is expressed as a general expression describing the energy scales forming smaller and larger matter entities in the universe. The energy will thus be given by;

$$\eta c + \beta c R_s + \mu c R_s^3 + \ldots\ldots\ldots + \delta c R_s^{N-1} = n\hbar c/R_s \;[2]$$

Note: the background space described by the Schwarzschild radius has changed, thus the above equation in any case can be used to calculate the basic properties of Black holes. Remember the Schwarzschild radius is the radius for a given mass where, if that mass could be compressed to fit within that radius, no known force or degeneracy pressure could stop it from continuing to collapse into a gravitational singularity[34].

The mass equation

Having explored the energy scale we now form general equation that describes well the mass scale. This is also done the same way as eqn2 and therefore generate,

$$\eta/c + \beta R_s/c + \mu R_s^3/c + \ldots\ldots\ldots + \delta R_s^{N-1}/c = n\hbar/cR_s \;[3]$$

The maximal magnetic field

Assuming that the energy $W = \beta c R_s$, from eqn2 is equal to the energy $W = mc^2$, we hence obtain the magnetic field as, $B = c^3/8\pi Ge = 1.0054\times10^{53}$ N/Am. using this magnetic field in the energy equation, $W = \eta c$ we get the energy in the

---

form $W = (c^2/2)\sqrt{\hbar c/G}$ where the quantity $\sqrt{\hbar c/G}$ is the Planck mass M_P at an energy of 6.119×10^{18} GeV.

Time taken by a black hole to evaporate and its entropy

The energy required here is given in Eqn2, it is at this, that the intensity $P = W/A\Delta t$, (where A is the area and t is the time) is used. We take the energy $W = \mu c R_s^3$ (from Eq2) as our interest from which we obtain the time as $\Delta t = 256\pi^3 R_s^3/Ac$. But with black holes the area will become exactly equal to the square of the Planck length as $A \sim L^2_p = \hbar G/8\pi\, c^3$ hence the change in time is given by $\Delta t = 63500.86\pi\, G^3 m^3/\hbar\, c^4$.

For entropy we set the energy to kT, where k is Stefan's-Boltzmann's constant and T is the temperature of the body. Now for $kT = \mu c R_s^3$, since Δt is known the entropy is thus given by $S = W/T = 78.96Ak\, c^3/\pi\hbar G \sim A/4$. In conclusion we state that the entropy of a black hole is proportional to the area of the event horizon.

The quantum Hall Effect

For this effect the momentum η is used. From Eqn2 we set, $\eta c = n\hbar / R_s$ which gives the magnetic flux as $4\pi R_s^2 B = nh/e$, from which the resistance is given by $\zeta = 4\pi R_s^2 B /e = nh/e^2$. for n= 1,2,3,4 the resistance is of a value 25833.8Ω.

Maximum Intensity

Using eqn3 in this case, since B is known and P got from $\mu R_s^4 = n\hbar$; as $P = \hbar c^2/256\pi^3 R_s^4$, we hence obtain, $M_P/2 + m + M_P/m = M_P/m$, which gives $M_P + 2m = 0$, and for identical mass M =0, which is true. The intensity at the planck length that is for $R_s = L_p$ is

$$P = c^8/\pi\hbar G^2$$

CHAPTER7: On The Complete Theory of Light

Basing our study on the electric currents generated whenever there is a changing magnetic field (B) and a changing electric field (E) in the electromagnetic wave we can construct a complete theory for the electromagnetic radiations. The theory is created using the symmetry between a long wire placed in the electromagnetic fields which induce vibrating electrons that carry current in the wire and the electromagnetic wave which constitute changing electric and magnetic fields that create vibrating photons in the wave. Therefore a wire is to a wave what a vibrating electron is to a vibrating photon in the wire and a wave respectively. The aim of the paper is to give a clear description of the theory of electromagnetic radiations (light). The goal of the paper on the other hand is to show that the wave-particle descriptions of reality can be applied to any physical situation simultaneously. The objective of the paper is to show that the Photoelectric Effect and the Compton Effect can both be explained by the wave model and the particle model at the same time.

Consider a long wire connected to an ammeter and strong electric and magnetic fields produced in a vacuum. Let us assume that whenever a wire is brought in vicinity of a changing electric field, electrons of mass (m) are set into motion in the wire and then an ammeter deflects, recording a current (i_E). The current in the wire due to a changing electric field should be given by

$$i_E = \frac{j\varepsilon_0}{2\pi m}E \quad (1)$$

Where (ε_0) is the permittivity of free space and (j) is the constant of action in SI units Js. therefore the current is quantized and depends on both the electric field and the mass of an electron.

When the wire is brought into the magnetic field, vibrating electrons at a frequency of oscillation (f) are set in motion at a speed (v) through the wire generating a current given by

$$i_B = \frac{v}{2\pi\mu_0 f}B \quad (2)$$

Where (μ_0) is the permeability of free space.

Assuming that the ammeter records different values of (i_E) and (i_B), what will be the change in the current values recorded at the ammeter? Subtracting equation (1) from equation (2) we have

$$\Delta I = (i_E - i_B) = \left(\frac{j\varepsilon_0}{2\pi m} E - \frac{v}{2\pi \mu_0 f} B \right) \tag{3}$$

This is the change in the currents due to changing magnetic and electric fields. Assuming that there is no change in the current, meaning that the current values for i_E are equal to those of i_B (i.e $\Delta I = 0$). This will imply that the magnetic field strength was equal to the electric field strength at one point in both experiments. In terms of electromagnetic radiations in the vacuum, assuming that a wire carrying current is replaced by a wave and electrons are replaced by photons. The wire replaced by a wave is made up of vibrating electric and magnetic fields at a given frequency making an electromagnetic wave. The electrons replaced with photons will represent the particle properties of the electromagnetic wave (light) with associated mass and speed (v).

The symmetry here is between the long wire and the wave, the electrons and the Photons. The electric and magnetic fields brought in vicinity of the wire and the number of oscillations per second of the electron in the wire is what leads to an electromagnetic wave. The electrons with a given mass and moving at a given speed is what constitute a photon. Then at $\Delta I = 0$, we have on arranging,

$$\frac{jf}{mv} = \frac{1}{2\pi \mu_0 \varepsilon_0} \frac{B}{E} \tag{4}$$

This means that at $\Delta I = 0$, either a changing magnetic field or a changing electric field produces a current. Then it should be true that a changing magnetic field produces an electric field just as a changing electric field produces a magnetic field. This process in the electromagnetic wave continues indefinitely. The electromagnetic wave will move at a constant speed (c), since for electromagnetic waves, $\frac{E}{B} = c$, and for a photon $\frac{jf}{mv} = c$ where j=6.63× 10^{-34}Js (also called the Planck constant after Max Planck) and mv is the photon momentum. Implying that the photon energy is related to the frequency of the electromagnetic wave by (jf). Then the electromagnetic wave will move

at a constant speed given as, since by symmetry $\frac{E}{B} = \frac{jf}{mv} = c$

$$c = \frac{1}{\sqrt{\varepsilon_o \mu_o}} = 2.99792458 \times 10^8 \, \frac{m}{s}$$

Where $\varepsilon_o = 8.85418782 \times 10^{-12} \, \frac{c^2}{Nm^2}$ and

$$\mu_o = 1.26 \times 10^{-6} \frac{Ns^2}{c^2}$$

We have therefore deduced based on the symmetry between a current (electron) carrying wire in the electromagnetic field and the photons in electromagnetic waves that an electromagnetic wave moves at a constant speed of light. It is also true from the deductions that light is indeed made up of particles of light called photons and vibrating electric and magnetic fields. The deduction would not be possible if the wave and particle descriptions of the situations had not been applied simultaneously (into what is called "the wave-particle duality).

Unexpectedly enough the **photoelectric effect** can also be explained by Equation (3), on arranging

$$\frac{2\pi mf}{\varepsilon_o E} \Delta I = jf - \frac{mv}{2\pi \mu_o \varepsilon_o} \frac{B}{E}$$

Then the total energy of the particle of light (Photon) is then given by

$$jf = \frac{2\pi mf}{\varepsilon_o E} \Delta I + \frac{mv}{2\pi \mu_o \varepsilon_o} \frac{B}{E} \quad (5)$$

It is therefore true that the photoelectric effect can be explained when both the particle and wave models of reality are applied in the experiment at the same time (simultaneously). The work function from Einstein's photoelectric equation (A. Einstein, 1905) will here be replaced by $\frac{2\pi mf}{\varepsilon_o E} \Delta I$ while the kinetic energy of the electrons at the surface of the metal will be given by $\frac{mv}{2\pi \mu_o \varepsilon_o} \frac{B}{E}$. Equation (5) reduces to Einstein's Photoelectric effect when, the speed of the electron is $v = \frac{1}{\pi \mu_o \varepsilon_o} \frac{B}{E}$ and the change in current for a complete circuit is $\Delta I = \frac{j\varepsilon_o E}{2\pi m}$.

The validity of the **Compton Effect** can also be deduced from Equation (3). The current can be taken as the product of the frequency (f) of radiations and the charge (q) on the particle. Then the current due to the electric field is $i_E = qf_1$ and that due to the magnetic field is $i_B = qf_2$. In the case of the Compton Effect, q is the charge on the free electron while f_1 and f_2 are the frequencies of the incoming photon and outgoing photon after collision with the free electron respectively. Then equation (3) can be written as

$$f_1 - f_2 = \frac{1}{q}\left(\frac{j\varepsilon_0}{2\pi m}E - \frac{v}{2\pi \mu_0 f}B\right) \quad (6)$$

Since photons move with the speed of light(c) then their frequencies is related to their speed and wavelength by $f = \frac{c}{\lambda}$, then we have

$$\frac{1}{\lambda_1} - \frac{1}{\lambda_2} = \frac{1}{qc}\left(\frac{j\varepsilon_0}{2\pi m}E - \frac{v}{2\pi \mu_0 f}B\right)$$

On arranging to include the charge density of the free electron for electric field lines in an area of $\frac{\lambda_1 \lambda_2}{2\pi}$, we obtain

$$\frac{2\pi q}{\lambda_1 \lambda_2}(\lambda_2 - \lambda_1) = \frac{j}{mc}\left(\varepsilon_0 E - \frac{mv}{\mu_0 jf}B\right)$$

Where (mc) is the momentum of an electron treated relativistic ally, on letting the charge density $= \frac{2\pi q}{\lambda_1 \lambda_2} = \varepsilon_0 E$, we deduce the change in the wave length of the incoming photon and outgoing photon after collision with the free electron as

$$\Delta\lambda = (\lambda_2 - \lambda_1) = \frac{j}{mc}\left(1 - \frac{mv}{\rho\mu_0 jf}B\right)$$

Since $\rho = \varepsilon_0 E$, we then have

$$\Delta\lambda = (\lambda_2 - \lambda_1) = \frac{j}{mc}\left(1 - \frac{\frac{mvB}{\varepsilon_0\mu_0 E}}{jf}\right)$$

Since jf is the energy carried by the photon, and then also $\frac{mvB}{\varepsilon_0\mu_0 E}$ is the energy carried by the free electron. Treating the electron relativistically such that for electromagnetic waves moving at a speed (v) relative to the electron moving at a speed of light $c = \frac{1}{\sqrt{\varepsilon_0\mu_0}}$, the electric field in the wave will be related to the magnetic field by $Bv = E.$ then the energy carried by an electron can be given by mc^2. Then the angle at which the photon is scattered after collision with the free electron will be given by

$$\theta = \cos^{-1}\left.\left(\frac{mvB}{\varepsilon_0\mu_0E}\right)\right/jf$$

$$(7)$$

Where mv is the momentum of the photon in the electromagnetic wave consisting of a changing electric field E and magnetic field B both moving at a constant speed of light $c = \frac{1}{\sqrt{\varepsilon_0\mu_0}}$. Treating the electron relativistically we have

$$\theta = \cos^{-1}\frac{mc^2}{jf}$$

When the energy carried by the photon is equal to the energy possessed by the electron then $\theta = 0$, meaning that there is or there is no scattering and whatsoever there is no increase in photon wavelength hence $\Delta\lambda = 0$.

A complete theory of light can't fail to explain **the structure of an atom**. I therefore take a complete discussion of what goes on inside an atom only with the help of Bohr's energy levels which he derived using classical mechanics and quantum theory. Let $\Delta f = f_1 - f_2$ be an increase in the frequency of the electromagnetic radiations emitted from an atom. Then squaring both sides of equation (6) and arranging will give

$$4\pi^2\Delta f^2 = \frac{1}{m^2q^2}\left(j\varepsilon_0 E - \frac{mv}{\mu_0 f}B\right)^2$$

$$4\pi^2 m^2 q^2 \Delta f^2 = j^2\varepsilon_0^2 E^2 - 2\frac{j\varepsilon_0 EBmv}{\mu_0 f} + \frac{B^2 m^2 v^2}{\mu_0^2 f^2}$$

Dividing through by $64\pi^4 j^2 \varepsilon_0^2$ and multiplying through by q^2 gives the energy of the atom as on arranging

$$\frac{mq^4}{16\pi^2\pi^4 j^2 \varepsilon_0^2} = \frac{1}{64\pi^4 m\Delta f^2}\left((Eq)^2 - 2\frac{m(Eq)(Bqv)}{\mu_0\varepsilon_0(jf)} + \frac{(Bqv)^2 m^2}{\mu_0^2\varepsilon_0^2(jf)^2}\right)$$

The energy of the n-th level is since the reduced Planck constant is $n\hbar = \frac{nj}{2\pi}$

$$\frac{mq^4}{32\pi^2\pi^4 n^2 \hbar^2 \varepsilon_0^2} = \frac{1}{32\pi^2 n^2 m\Delta f^2}\left((Eq)^2 - 2\frac{m(Eq)(Bqv)}{\mu_0\varepsilon_0(jf)} + \frac{(Bqv)^2 m^2}{\mu_0^2\varepsilon_0^2(jf)^2}\right)$$

The expression on the left hand side of the equation is the quantized energy of an atom (Niels Bohr, 1913) while the right hand side of the equation represents the energy of the atom in terms of the forces associated with it. In the equation we let $H_e = Eq$ be the electric force for a particle moving in the electric field and $H_b = Bqv$, the magnetic force on a particle with charge q moving in the magnetic field. Since the speed of light is $= \frac{1}{\sqrt{\varepsilon_0\mu_0}}$, then the quantized energy can be given as

$$W_n = \frac{1}{32\pi^2 n^2 m\Delta f^2}\left(H_e^2 - 2\frac{H_e H_b mc^2}{jf} + \frac{H_b^2(mc^2)^2}{(jf)^2}\right)$$

Then on arranging we obtain

$$W_n = \frac{1}{32\pi^2 n^2 m\Delta f^2}\left(H_e - \frac{mc^2}{if}H_b\right)^2 \quad (8)$$

When the energy of an electron moving at a speed of light in atom is equal to the energy of the emitted photon, then

$$W_n = \frac{1}{32\pi^2 n^2 m\Delta f^2}(H_e - H_b)^2 = \frac{1}{32\pi^2 mn^2}\left(\frac{\Delta H}{\Delta f}\right)^2 \quad (9)$$

Where $\Delta H = H_e - H_b$ is the difference or change between the electric force and the magnetic force in an atom, when the two forces balance (i.e. $H_e = H_b$), then $W_n = 0$ meaning that the total energy of an atom will cease to exist.

Therefore the total energy of an atom increases with the square of the change in the electric and magnetic forces which govern an electron but falls off as the square of the change in the frequency of the radiation emitted by it.

From equation (8) the ratio of the energy of an electron to that of the photon $\frac{mc^2}{jf}$, is the limit at which if the energies are not equal you will not get a change in the electric and magnetic forces. Treating the ratio as a number $\tau = \frac{mc^2}{jf}$, we get from equation (8)

$$W_n = \frac{1}{32\pi^2 mn^2}\left(\frac{H_e - \tau H_b}{f_1 - f_2}\right)^2 \quad (10)$$

When $\tau = 0$, it means that the relativistic energy (mc^2) of an electron in an atom is zero, and that the total energy of an atom only increases with the electric force on the electron. The relationship (equation 10) is a complete expression for the laws according to which, by the theory here advanced, the structure of an atom should be viewed.

In conclusion, a complete theory of light is only possible if both the wave and particle descriptions of reality are applied to the physical situation at the same time. In discussing Young's double slit experiment for example we should be able with the formulas given above to treat the electromagnetic radiations on both a wave and particle model.

CHAPTER8: Construction of a Consistent Physical Theory of Nature

A consistent theory of nature, simply the "theory of everything" is constructed using one of the profound ideas of "axioms", that, when the Stoney units of measure are multiplied by the coupling constant (a dimensionless number) of a form $\alpha^{\frac{n-1}{2}}$, one can easily calculate the mass of all particles in the universe and their length or time scales with accuracy provided the value of n is known. The mass of the electron is calculated at $\alpha=1/137.036$ and n=21.32 while the mass of the earth is calculated at n= -29.99, hence solving one of the major unsolved problems in physics. The Planck mass is calculated and determined in principle to be 5.4556×10^{-8} Kg, a different value from the given value would lead to variations in our fundamental physical constants of electricity and gravity. The energy scales at given length scales in literature are also deduced in which a requirement to revisit our profound known physical theories is proposed.

Introduction

One of the major unsolved problems in physics is developing a final theory, ultimate theory or theory of everything. In this paper we present a series of hypotheses and speculations leading inescapably to a conclusion that when the Stoney fundamental units of measure are multiplied by the electromagnetic coupling constant (fine structure constant) powered by any integer, $\alpha^{\frac{n-1}{2}}$ from 0,1,2,..................,n, one gets to calculate the mass of all particles in the universe, the lengths between them and the time expressible at a scale of the known fundamental physical constants of nature. Our hypotheses may be wrong and our speculations idle, but the uniqueness and simplicity of our scheme are reasons enough that it be taken seriously.

Our starting point is the assumption that all of the fundamental physical units of measure can be calculated and organized to demonstrate different branches and scales of physics whatsoever using the following formulas,

<u>Length</u>

$$L_n = \frac{e}{c^2}\sqrt{\frac{G}{2\varepsilon_0}}\,\alpha^{n-1} \tag{1}$$

Time

$$t_n = \frac{e}{c^3}\sqrt{\frac{G}{2\varepsilon_0}}\,\alpha^{n-1} \tag{2}$$

Mass

$$M_n = e\sqrt{\frac{1}{2G\varepsilon_0}}\,\alpha^{n-1} \tag{3}$$

Where α is the coupling constant for either electromagnetic or gravitational interactions, G is the universal gravitational constant, e is the elementary charge on an electron, c is the speed of light and ε_0 is the permittivity of free space, the meaning of n is left to be investigated as per the meaning of the theory.

CASE1:

We derive the fundamental units of measure at values of n=0, 1,2,3,4 and 5 only for the fine structure constant $\alpha = \frac{e^2}{4\pi\varepsilon_0\hbar c}$ where is the reduced Planck constant $\hbar = \frac{h}{2\pi}$.

At n=0

$$L_o = \sqrt{\frac{2\pi G\hbar}{c^3}}, \quad t_o = \sqrt{\frac{2\pi G\hbar}{c^5}}, \quad M_o = \sqrt{\frac{2\pi\hbar c}{G}}$$

At n=1

$$L_1 = \frac{e}{c^2}\sqrt{\frac{G}{2\varepsilon_0}}, \quad t_1 = \frac{e}{c^3}\sqrt{\frac{G}{2\varepsilon_0}}, \quad M_1 = \frac{e}{\sqrt{2G\varepsilon_0}}$$

At n=2

$$L_2 = \frac{e^2}{\varepsilon_0}\sqrt{\frac{G}{8\pi c^5\hbar}}, \quad t_2 = \frac{e^2}{\varepsilon_0}\sqrt{\frac{G}{8\pi c^7\hbar}}, \quad M_2 = \frac{e^2}{\varepsilon_0}\sqrt{\frac{1}{8\pi G\hbar c}}$$

At n=3

$$L_3 = \frac{e^3}{\pi c^3}\sqrt{\frac{G}{32\varepsilon_0{}^3}}, \quad t_3 = \frac{e^3}{\pi c^4}\sqrt{\frac{G}{32\varepsilon_0{}^3}}, \quad M_3 = \frac{e^3}{\pi\hbar c}\sqrt{\frac{1}{32\varepsilon_0{}^3 G}}$$

At n=4

$$L_4 = \frac{e^4}{\varepsilon_0^2}\sqrt{\frac{G}{128\pi^3\hbar^3c^7}},$$

$$t_3 = \frac{e^4}{\varepsilon_0^2}\sqrt{\frac{G}{128\pi^3\hbar^3c^9}},$$

$$M_3 = \frac{e^4}{\varepsilon_0^2}\sqrt{\frac{1}{128\pi^3 G\hbar c}}$$

At n=5

$$L_5 = \frac{e^5}{\pi^2\hbar^2c^4}\sqrt{\frac{G}{512\varepsilon_0^5}},$$

$$t_5 = \frac{e^5}{\pi^2\hbar^2c^5}\sqrt{\frac{G}{512\varepsilon_0^5}},$$

$$M_3 = \frac{e^5}{\pi^2\hbar^2c^2}\sqrt{\frac{1}{512G\varepsilon_0^5}}$$

At n=0, we obtain the Planck natural units while at n=1, we obtain the Stoney units of measure

CASE2:

We further derive the fundamental units of measure at values of n=0, 1, 2, only for the gravitational coupling $\alpha = \frac{Gm^2}{\hbar c}$

At n=0

$$L_0 = \frac{e}{m}\sqrt{\frac{\hbar}{2\varepsilon_0 c^3}},$$

$$t_0 = \frac{e}{m}\sqrt{\frac{\hbar}{2\varepsilon_0 c^5}},$$

$$M_0 = \frac{e}{Gm}\sqrt{\frac{\hbar c}{2\varepsilon_0}}$$

At n=1

$$L_1 = \frac{e}{c^2}\sqrt{\frac{G}{2\varepsilon_0}},$$

$$t_1 = \frac{e}{c^3}\sqrt{\frac{G}{2\varepsilon_0}},$$

$$M_1 = \frac{e}{\sqrt{2G\varepsilon_0}}$$

At n=2

$$L_2 = meG\sqrt{\frac{1}{2\varepsilon_0 c^5\hbar}},$$

$$t_2 = meG\sqrt{\frac{1}{2\varepsilon_0 c^7\hbar}},$$

$$M_2 = em\sqrt{\frac{1}{2\varepsilon_0\hbar c}}$$

It proves difficult to deduce the Planck units here, simply because the charge and mass do not cancel out. But if you set the ratio of charge to mass at n=0 in the above formulas as $\frac{e}{m} = \sqrt{4\pi\varepsilon_0 G}$, one obtains the Planck units. Also, one obtains the values of n=2 in Case1 when we substitute for $m = \frac{e}{\sqrt{4\pi\varepsilon_0 G}}$, in Case2, for n=2. This means that, the formulas which do not exist in case2 but are present in case1 can be calculated by applying a simple formula, $e = m\sqrt{4\pi\varepsilon_0 G}$ and vise versa is true.

When we substitute for $e = m\sqrt{4\pi\varepsilon_0 G}$, at n=1, we obtain,

$$L_1 = \frac{Gm}{c^2}\sqrt{2\pi}, \quad t_1 = \frac{Gm}{c^3}\sqrt{2\pi}, \quad M_1 = m\sqrt{2\pi}$$

This represents formulae at a scale of general relativity, in which it is deduced here that, the mass of a particle in both the special and general relativity theory makes sense when multiplied by a constant $\sqrt{2\pi}$.

When $me = m^2\sqrt{4\pi\varepsilon_0 G}$ at n=2 above, we obtain

$$L_2 = m^2\sqrt{\frac{2\pi G^3}{c^5\hbar}}, \quad t_2 = m^2\sqrt{\frac{2\pi G^3}{c^7\hbar}}, \quad M_2 = m^2\sqrt{\frac{2\pi G}{\hbar c}} = \frac{2\pi m^2}{m_p}$$

Where m_p is the Planck mass $\sqrt{\frac{2\pi\hbar c}{G}}$

It should however be taken seriously from the above investigation that changing the number 2π in the formulas (case1, at n=0, the planck units/scales), will change the statement of the formula $e = m\sqrt{4\pi\varepsilon_0 G}$, which will mean that the values of the fundamental physical constants $\frac{1}{4\pi\varepsilon_0}$,G are varying, therefore in order to maintain the constants unchanged we have to maintain the Planck units unchanged in formula as they are derived here. Thus the Planck mass will have a mass given by, $2.2176470119 \times 10^{-8}\sqrt{2\pi}$ =5.4556 $\times 10^{-8}$ Kg.

At present there is no candidate theory of everything that includes the standard model of particle physics and general relativity. For example, no candidate theory is able to calculate the mass of an electron. However in this paper the mass of an electron is deduced when n=21.32 and α=1/137.036 as,

$$M_{21.32} = e\sqrt{\frac{1}{2G\varepsilon_0}}\left(\frac{1}{137.036}\right)^{20.32} =$$

$9.082073363 \times 10^{-31} kg$

$$L_{21.32} = 6.745125 \times 10^{-58} m$$

$$t_{21.32} = 2.25 \times 10^{-66} s$$

Also the proton mass is deduced at n= 18.26, and $\alpha=1/137.036$ as,

$$M_{18.26} = e\sqrt{\frac{1}{2G\varepsilon_0}}\left(\frac{1}{137.036}\right)^{17.26} =$$

$$1.688659377 \times 10^{-27} kg$$

Other masses including the mass of the earth (n= -29.99) can be deduced in the same way. It is important to note that, the value of n is negative for massive particles (e.g mass of the Sun and earth) but positive for microscopic particles like electrons and protons.

It is hereby noted that the values of the energy scales corresponding to the given length scale are off the scale and do not necessarily represent phenomenon at each given length scale. The values of these energy scales for each interaction as quoted in scientific literature will prove to be different from the ones represented here.

For example;

(1) For atomic length scale with $l_a \sim 10^{-10} m$, the value of n to be used in calculating other scales will be given by n= -22.83, from which the energy scale can be calculated as, $E_{-22.83} = M_{-22.83} c^2 \sim 7.6 \times 10^{43} GeV$

(2) For strong interaction length scale with $l_s \sim 10^{-15} m$, the value of n to be used in calculating other scales will be given by n= -18.151, from which the energy scale can be calculated as, $E_{-18.151} \sim 7.6 \times 10^{38} GeV$

(3) For electroweak interaction length scale with $l_w \sim 10^{-18} m$, the value of n to be used in calculating other scales will be given by n= -15.343, from which the energy scale can be calculated as, $E_{-15.343} \sim 7.6 \times 10^{35} GeV$

Discussion

It is possible that the correct theory of everything has been found in the formulas given above. It therefore seems appropriate for the reader or researchers to consider the calculation and determination of the values of the masses of all particles at given length/time scales in the universe with explicit accuracy, even if some people may consider such an enterprise premature or foolhardy. It is worth noting that the length and time scale through which one can probe the whole mass of the earth is $4.439 \times 10^{-3} m$ and $1.4806 \times 10^{-11} s$ respectively. It is therefore important to know how one can calculate the mass of any particle with accuracy and then inquire with simplicity into the length and time scale at which such a particle can be studied. This then means that any consistent theory of nature like the one constructed would be able to deduce the required derived quantities (i.e. voltage, current, magnetic field etc) from the given formulas for fundamental physical units of measure of mass, length and time without a need to inquire into other theories like the standard model, string theory or quantum gravity

CHAPTER9: The Basic Wave Equation (Derivation of the Fermi's Energy)

Consider a relativistic particle such that its mass changes with the electromagnetic fields. Arranging Eqn (6) to get the ratio of the relative speed of the particle to that of light

$v/c = (AE\ e\ /2nhc)$

According to special relativity the ratio of the speeds are in the form

$V/c = \sqrt{(1 - m_o^2/m^2)}$

Equating the two equations and multiplying through by c^2 we get

$$E^2 - E_o^2 = \frac{m^2 c^2 k^2 e^4}{n\hbar^2} = EE_n$$

Where E is the total relativistic energy of the particle (mc^2) and E_o is it's rest energy. Multiplying both sides of the eqn by $2m/\hbar$ and taking into account that

$E_n = mk^2 e^4 / 2n\ \hbar^2$

We have the square of the wave length as

$$\left[\frac{h}{mc}\right]^2 = \lambda^2 = \frac{\hbar^2}{2m(E - E_0)}\left[\frac{16\pi^2 E_n}{E + E_0}\right]$$

Where En is the energy of the quantized state of the hydrogen atom, this is the simplest form of the **Schrödinger equation.**

When E_o=0 and $E_n = E/4$ the momentum of the non relativistic particle is given by

$p = \sqrt{2mE}$

This is also called the Fermi's momentum where E is the Fermi's energy.

References

(1)Balungi Francis, (2010) "A hypothetical investigation into the realm of the microscopic and macroscopic universes beyond the standard model" general physics arXiv:1002.2287v1[1] [physics.gen-ph]

(2) Hawking, Stephen[2] (1975). "Particle Creation by Black Holes"[3]. Commun. Math. Phys.[4] 43 (3): 199–220. Bibcode[5]:1975CMaPh..43..199H[6].

(3)Hawking, S. W.[7] (1974). "Black hole explosions?". Nature.248(5443):30–31.

Bibcode[8]:1974Natur.248...30H[9].doi[10]:10.1038/248030a0[11].

(4)Carlo Rovelli (2003) "Quantum Gravity" Draft of the Book Pdf

(5)Some few texts used are from Wikipedia https://creativecommons.org/licenses/by-sa/3.0/

---

1. https://arxiv.org/abs/1002.2287v1

2. https://en.wikipedia.org/wiki/Stephen_Hawking

3. http://www.springerlink.com/content/c4553033029k5wk6/

4. https://en.wikipedia.org/wiki/Commun._Math._Phys.

5. https://en.wikipedia.org/wiki/Bibcode

6. http://adsabs.harvard.edu/abs/1975CMaPh..43..199H

7. https://en.wikipedia.org/wiki/Stephen_Hawking

8. https://en.wikipedia.org/wiki/Bibcode

9. http://adsabs.harvard.edu/abs/1974Natur.248...30H

10. https://en.wikipedia.org/wiki/Digital_object_identifier

11. https://doi.org/10.1038%2F248030a0

Thank you very much for buying my book.

You have supported my research works towards the unification of gravity with quantum mechanics.

Connect with Balungi Francis

I really appreciate you reading my book! Here are my social media coordinates:

Friend me on Facebook: http://facebook.com/balungi.francis[12]

Visit my Facebook page Visionary School of Quantum Gravity: http://facebook.com/BalungiF[13]

Subscribe to my blog: http://visionaryschoolofquantumgravity.blogspot.ug[14]

Contacts, Tel: +256(0)777105605

+256(0)703683756

Email Address: balungif@gmail.com

---

12.   http://facebook.com/markcoker

13.   http://facebook.com/markcoker

14.   http://blog.smashwords.com

Don't miss out!

Visit the website below and you can sign up to receive emails whenever Balungi Francis publishes a new book. There's no charge and no obligation.

https://books2read.com/r/B-A-LFLG-EXOT

BOOKS 2 READ

Connecting independent readers to independent writers.

Also by Balungi Francis

Beyond Einstein
Quantum Gravity in a Nutshell1
Solutions to the Unsolved Physics Problems
Mathematical Foundation of the Quantum Theory of Gravity
A New Approach to Quantum Gravity
Balungi's Approach to Quantum Gravity
QG: The strange theory of Space,Time and Matter
The Holy Grail of Modern Physics
Fifty Formulas that Changed the World
Quantum Gravity in a Nutshell1 Second Edition
What is Real?:Space Time Singularities or Quantum Black Holes?Dark Matter
or Planck Mass Particles? General Relativity or Quantum Gravity? Volume or
Area Entropy Law?
The Holy Grail of Modern Physics
Brief Solutions to the Big Problems in Physics, Astrophysics and Cosmology

Brief Solutions to the Big Problems
Brief Solutions to the Big Problems

Pursuing a Nobel Prize
Serious Scientific Answers to Millennium Physics Questions

Using Geographical Information Systems to Create an Agroclimatic Zone map
for Soroti District

Think Physics
Proof of the Proton Radius
Emergence of Gravity
On the Deflection of Light in the Sun's Gravitational Field
Reinventing Gravity

Standalone
Using Gis to Create an Agroclimatic Zone map for Soroti Distric
Expecting
Quantum Gravity in a Nutshell 2
Balungi's Guide to a Healthy Pregnancy
Prove Physics
The Origin of Gravity and the Laws of Physics
Derivation of Newton's Law of Gravitation
When Gravity Breaks Down

www.ingramcontent.com/pod-product-compliance
Lightning Source LLC
Chambersburg PA
CBHW030019190526

45157CB00016B/3136